MAKING IT HAPPEN

RALPH MENZANO

MAKING IT HAPPEN
. . . DOESN'T HAVE TO BE HARD
BY RALPH MENZANO

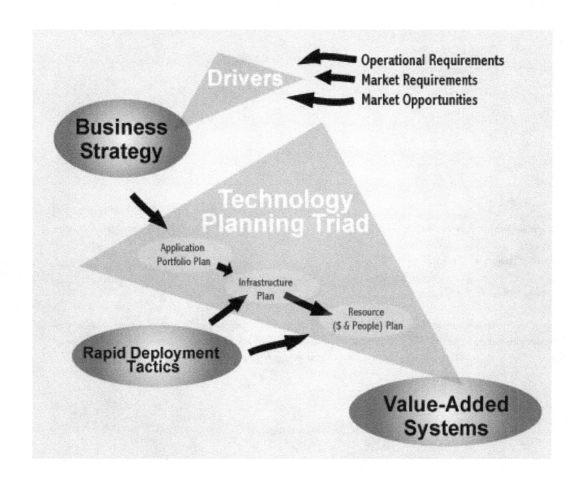

authorHOUSE®

AuthorHouse™
1663 Liberty Drive
Bloomington, IN 47403
www.authorhouse.com
Phone: 1-800-839-8640

First published by Publications Unbound, 2003
Republished by AuthorHouse, 9/13/2011

ISBN: 978-1-4670-3534-7 (sc)
ISBN: 978-1-4670-3535-4 (e)

Library of Congress Control Number: 2011916462

MAKING IT HAPPEN
. . . Doesn't Have To Be Hard

BY RALPH MENZANO

DEDICATION

TO MY WIFE ALICE AND MY DAUGHTERS KELLEY AND STEFANIE WHO TOGETHER CONTINUALLY INSPIRE ME.

ABOUT THE AUTHOR

Ralph Menzano has been in the computer field for 25 years - from when it was called "Data Processing" in the '70s, to "Management Information Systems" in the '80s, to its more recent metamorphosis as "Information Technology," or IT, for short. He has seen the industry go from punched cards to the Internet. As an active practitioner, i.e. installer of systems, Mr. Menzano had the unique opportunity to view the growth of IT in three distinct industries - manufacturing, finance, and government.

Ralph Menzano is currently the Executive Director for Oracle Corporation's Transportation Industry Solutions, encompassing highway authorities, toll ways, airports, seaports, and transit agencies.

Mr. Menzano has held the position of:

- Chief Information Officer (CIO) of Southeastern Pennsylvania Transportation Authority (SEPTA), Philadelphia, PA, a $1.5 billion Transit Agency (5th largest in the U.S.). In this capacity, he established a strategic technology plan including project priorities, technical infrastructure, and resource requirements. In this capacity he managed/influenced over 400 systems valued at over $100 million.
- Vice President, Technology, The Chase Manhattan Bank, New York, NY where he established IT infrastructure that enabled the growth of a Commercial Loan Division from zero to $1,000,000,000.
- Vice President, Systems Development, GMAC Mortgage Corporation, Horsham, PA where he constructed technical system plans for all business entities included in the entire loan process - origination, underwriting, servicing, and pooling for resale.
- Director, Systems Development, St Gobain N.A., where he constructed system plans and implemented projects for this global manufacturing conglomerate.
- Senior IT Auditor, U.S. Navy and Marine Corps performing operational, financial, and IT audits throughout the U.S. and Europe.

In these roles, he has managed every IT project imaginable. Some examples follow:

- IT Strategic Planning corresponding to Strategic Business Plans
- Strategic Sourcing/Outsourcing
- Year 2000 Compliance
- Installation of Local and Wide Area Networks
- All aspects of financial system suites_(General Ledger, Budgeting, Accounts Receivable, Accounts Payable, Fixed Assets, Payroll, Benefits)

- Manufacturing Enterprise Systems (Customer Order Management, Activity-based Costing, Just-in-Time Inventory, Process Control, Process Scheduling, Product Delivery)
- Banking Enterprise Systems (Loan Origination, Underwriting, Servicing, Pooling, Real Estate Tracking)
- Transportation Enterprise Systems (Route Planning, Revenue Collection, Incident Tracking, Fleet Management, Material Management)
- Voice-response Systems
- Internet Sites
- Document Imaging Systems
- Geographic Information Systems
- Call Center Management

Over the years, Mr. Menzano has become a nationally prominent speaker and author on the topics of Strategic Planning, Total Quality Management, Performance Measurement, Business Process Re-engineering, as well as general Information Technology. He has spoken at a wide variety of forums including APTA (American Public Transportation Association), BOMA (Builder/Owner Management Association), World Vision Conference, Information Week's Executive Advisory Council, Real Estate Finance Association, Institute of International Research, Quality Assurance Institute, Computer Security Institute. He has been an active member of the Villanova University Advisory Board, LaSalle University Advisory Board, and the Amtrak Advisory Board. He won St. Gobain's International Management & Special Recognition Award, the U.S. Navy's Superior Performance Award, and was recognize in a feature story of CIO Magazine. He holds a MBA from Philadelphia University & BS from Villanova University, is a Certified Systems Professional (CSP), and has taught at the University of Pennsylvania. He can be reached most effectively via e-mail: ralph.menzano@oracle.com

Acknowledgements

As time passes, you experience ideas and methods for getting work accomplished. This set of experience in turn becomes ones own methodology, especially as an individual moves from job to job or, in my case, from industry to industry. The influences in my career have come bundled in the form of formal systems processes (such as Accenture's Method 1 or EDS' Systems Life Cycle) and in the form of supervisors who came to understand that "good people make good systems." Thus, I am compelled to acknowledge every boss that either gave me an order or asked that I solve an issue. It is their influence that allowed me to establish an understanding of how to implement IT in ANY industry. I also wish to acknowledge every one of the hundreds of employees within my departments who built and implemented systems to which we are proud to be associated.

I especially want to acknowledge Mr. Christopher Allen, currently with KPMG, a former supervisor at St. Gobain, and a lifelong friend who conducted a "peer review" of this manuscript prior to submission to the publisher.

Other references are cited as they occur in the text. Of these, the reader will observe a preponderance of quotes from materials of the Gartner Group, a leading source in the IT industry. Besides Gartner, I also gained a great deal of knowledge from IBM, Oracle, & Unisys Corporations over the years.

I wish to also acknowledge Mr. Robin Moore, my writing coach and distinguished author in his own right, who provided the insight I lacked in the mechanics of compiling a book and with it the discipline of format, and Mr. Steve Andriole, a life-long mentor whose work I've admired for many years and whose encouragement has been continuous.

A Word from the Author

In this book, I have tried to detail common themes of Information Technology (IT) work that can be viewed as running horizontally across all industries. Within these chapters, I will refer upon my experiences in Manufacturing, Banking, and Transportation industry sectors.

The best way to view this book in through the single diagram that appears below:

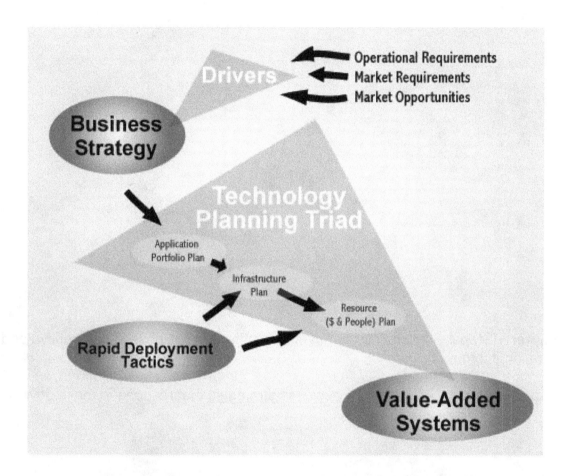

Each segment of this diagram is a chapter, and the chapters refer to significant, sequential steps needed to build an IT plan for any organization. Another way to view this methodology is:

A. Collaborative Effort

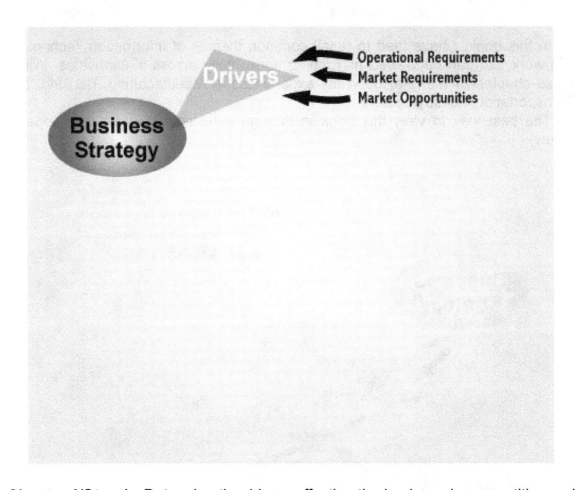

Chapter 1/Step 1 - Determine the <u>drivers</u> effecting the business in competitive and operational forms

Chapter 2/Step 2 - Ascertain the <u>business strategies</u> that respond to these drivers

B. Core Information Technology Responsibility Triad

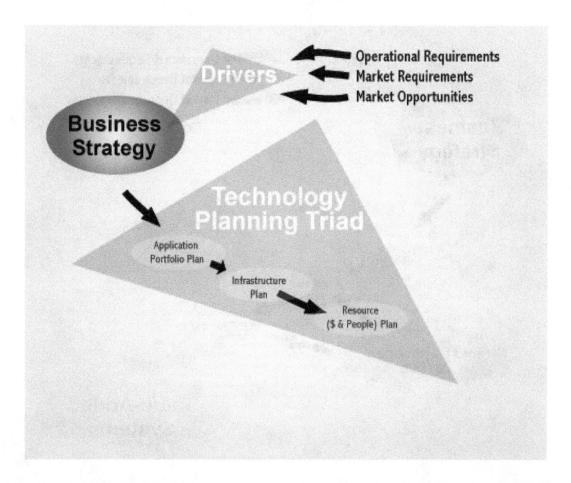

Chapter 3/Step 3 - Compile the list of IT <u>applications</u> needed to accomplish the business strategies

Chapter 4/Step 4 - Design the Technical <u>infrastructure</u> needed to support the applications

Chapter 5/Step 5 - Earmark the <u>money and staffing</u> needed to fulfill these IT projects

C. Collaborative Effort

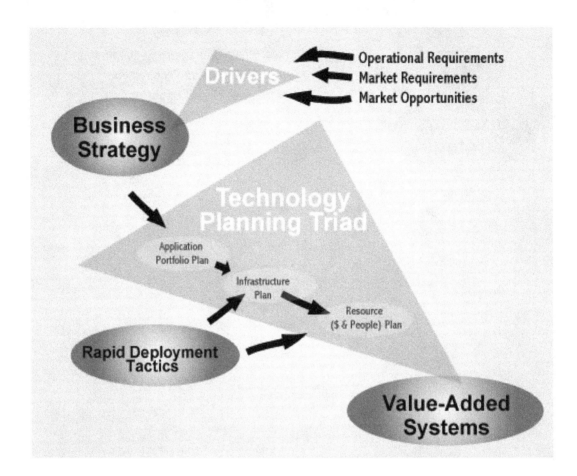

Chapter 6/Step 6 - Determine <u>shortcut options</u> to rapidly develop and deploy the applications and systems

Chapter 7/Step 7 - Implement Systems and integrate with processes/organizations where the entire company agrees the business can truly create <u>value</u>.

It is also important to understand that the sequence of the steps is essential, that is, businesses cannot start in the middle and hope to accomplish the desired results. A metaphor to think about is that of a successful crop. To receive a successful crop a farmer must think of an entire continuum:

- Water is created by winter thaw.
- Water flows freely from mountains through tributaries
- Channels are created to direct water where it can best nurture crops.
- Irrigation systems and techniques redistribute channeled water evenly and efficiently to field
- Fields are planted and harvested at optimal schedules
- Crops are delivered to markets.
- Value is returned to the farmer's overall business.

With complex information systems projects, the same sense of process exists. Companies cannot buy hardware before knowing that software they select fits their business requirements nor can they buy software before knowing the drivers of their business strategies. If the reader can understand this diagram, they understand the flow of this book, which in turn helps to understand how IT can be made simple, or at least be de-mystified.

A humorous story I often use to describe this stepwise flow is a baseball tale involving the New York Mets' Marv Thronberry. Marv was not an especially gifted player and was on the 1962 team that still holds the record for most losses in a single season. One day in Chicago, Marv hit a ball as well he could ever hit one. It looked like a home run off the bat but the fabled Chicago winds pushed the ball back into play causing Marv to run hard around the bases. When the outfielders didn't get the ball, he rumbled around first, chugged around second, and slid into third with what everyone thought was a triple. Just as the dust settled from the slide, Ernie Banks, the Chicago Cubs' first baseman called time out, asked for the ball, and touched first base. When the umpire signaled out, everyone realized that Marv failed to touch first base when he rounded the bases. The Mets' manager, Casey Stengel, came out of the dugout ready to start a rhubarb but Marv stopped him and said, "Casey, don't bother to argue about first base, because I missed second base too!"

To apply this lesson to IT projects, management needs to remember to touch all bases - understanding drivers and business strategies—before embarking on IT solution sets.

Lastly, and maybe most importantly, the book contains a series of questions triggered by the phrase:

"Ask Yourself"

With this trigger, the readers will make the subjects within the chapters more relative to their environment thereby allowing the reading of the book to become more interactive.

Chapter One

Drivers

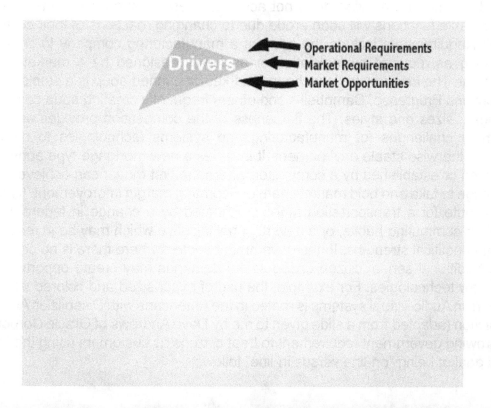

Organizations are driven in many directions in today's economy. They understand the mantra that ". . . standing still means falling behind . . ." but they also understand that a stable foundation is required. Speed to market for new products and services shares the focus with delivering today's products and services.

Ask Yourself:
What drives my industry, my company, my division(s), and my department)s)?

Drivers can be categorized into three groups - operational requirements, market requirements, and competitive leadership opportunities. Operational Requirements are those things that the firm must do to <u>maintain</u> current levels of service or product delivery. The ability of a manufacturing company to deliver its products as specified and on the date required; the ability of a banking institution to remit mortgage escrow funds to tax entities; or the ability of a transportation authority to meet it current route schedule. If a problem occurs with this stable set of functions, all efforts are diverted from reacting to market requirements and addressing competitive opportunities that seek to change to the business. So it is ironic that stability actually allows change to happen but, in the world of IT, it is an axiom.

Market Requirements are those projects instigated by change in the supply chain, by competitor reactions to customer preferences, or by legislation. In short, there is compelling need for change within the organization in order for that organization to compete. If the organization does not act upon these new requirements, it will find that its stable functions will soon erode due to changing reactions of their customers. Market Requirements include the need of a manufacturing company to create new packaging as required by legislation or as newly designed by a market-grabbing competitor. The low-growth, low-change market for canned soup is a stable industry, but contains Progresso, Campbell's and others frequently creating soup combination packages, sizes and styles. The fierceness of the competition provides an endless stream of challenges for manufacturing and systems technologies to meet in a mature, otherwise stable environment. It includes a new mortgage type approved by legislation or established by a competitor, where the first mover can achieve a terrific advantage to take and hold market share or operating margin improvement. It includes a new route for a transportation authority dictated by a change in federal funding, customer commuting habits, or a new regional initiative which may be in response to changing political strengths. In the government sectors where there is no competition in the traditional sense, citizen or regulatory demands may create opportunities for using new technologies. For example, the use of exact-sized and colored letters and numbers in Audio-visual systems is rooted in the Americans with Disabilities Act (ADA). A depiction (adapted from a slide given to me by Dave Andrews of Oracle Corporation) of a growing government requirement to treat citizens as customers using the Internet with a goal of being "on-line versus in-line" follows:

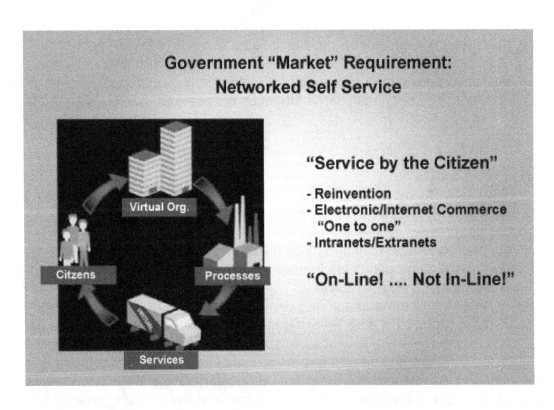

Competitive Opportunities include projects which are envisioned as gaining a distinct edge over others in the industry and to keep out new entrants. These are often initiated out of Marketing or Operational sides of the organization but occasionally are derived from the Finance side of the firm as ideas for achieving market leadership in cost, quality, value or logistics performance. These projects often involve enterprise-wide technologies and sweeping changes to the stable business model. Often, IT areas will first see the potential of some or all technologies at a strategic level, but Technology Department initiated projects are generally failures waiting to happen because they are not rooted within or sponsored by the profit centers of the firm.

The other point to note here is that Competitive Opportunities cascade downward, in that, over time, they become Market Requirements which then evolve into stable Operational Requirements. In some manufacturing organizations, activity-based cost accounting was once an Opportunity and now is an operational requirement. In banking, using the Internet was once novel for mortgage applications, but now it is expected. In Transportation, vehicle maintenance systems that were once a luxury are now a necessity to manage costs and operating risks.

Ask Yourself:
What drives a company to take on an IT project?

The answer to this question lies within the managerial visions, litanies of problems/ issues, customer hopes, board initiatives, new management directions, R&D efforts that come to fruition, changes for change sake, and other items of momentum that are often more intangible than tangible.

Examples of a Driver

U.S. Election uncertainties are driving governmental organizations around the country to improve their election processes. According to Gartner's December 2000 Monthly Research Review, the problems experienced by Florida (and by other states albeit less publicized) will heighten the case for electronic voting technology. Election.com performed the Arizona Democratic Primary (March 2000) - the first public election to feature the option to vote via the Internet. The election allowed citizens to cast their vote using four different means: a) Over the Internet during the 4-day period preceding the election; b) by mail until election day; c) via computer at the polls on election day; or d) using traditional paper ballots.

Arizona experienced some issues - servers were often unable to process the transaction volume, and the software was incompatible with Macintosh browsers and with a screen reader commonly used by the blind. The Voter help desk was flooded with calls. Nonetheless, about half of the 87,000 voters cast their vote electronically. The results of the voting were available 17 minutes after the poll closed. Residents who were out of state voted via Internet thereby eliminating the need for waiting for absentee ballots which can delay final tallies for weeks. Cost of manually cast ballots = $20. Gartner predicts that by the year 2004, all 50 states will be using some form of electronic voting. Any inconsistencies developing in a state without e-voting will generate a public outcry to get with it (and with the competition) and spend money on the new technology. It will not matter how much it costs.

The real driver in this case is the need for the United States to get more than 50 percent of its vote-capable public to actually cast votes. Other Drivers of governmental organizations are depicted in the following exhibit.

Government IT Drivers

Examples of items that are not drivers

Wireless Networks, Voice over the Internet, XML (Extensible Markup Language), and other new technologies are key to the growth and even survival of some firms, maybe even to a niche competitor of your firm. However, unless your company is in the technology industry, these represent opportunities that bear review, not a redirection of technology investment. For banking, manufacturing, government, transportation, retail, and other non-technology companies, technology represents a tool for business and NOT the business itself.

IT executives (and others) often treat opportunities *in general* as opportunities for a *specific business.* Just because some new thing is available or some competitors are using it doesn't mean that the same thing is an appropriate target for your company. It does mean that some thought should be given to opportunities to see if they align with the strategies, capabilities, and market positions held by the firm.

Drivers in Good Times or Bad Times

Bad Times have an effect on Drivers
- Economic Bad Times = diminished stock values; reduced funding; less than expected sales; need to reign in costs
- Operating Bad Times = lower voluntary turnover with limited opportunity to bring in new and different skills to maintain lead in innovation
- Industry Bad Times = lower prices, shorter product to market lead-times, over-capacity and emphasis on maintenance versus investment

Overall ROI for any project especially IT projects is more closely scrutinized than in good times. For example, IT organizations spent billions of dollars getting pass Y2K, a driver that was almost universally accepted by industries around the world. After the deadline, which many now claim was self-created by the technology sector, the return to traditional scrutiny levels was apparent.

Good Times or the anticipation of better results have an effect on Drivers

- Economic Good Times = rising stock values, increasing sales, opportunities to invest for competitive advantage
- Operating Good Times = higher turnover, new product pipeline, capacity investments, high potential for projects that are known to be 'doing the right thing' with strong cash flow
- Industry Good Times = higher demand, higher prices, higher opportunity to seize competitive advantage, i.e. can't throw a rock and miss the ocean.

In good economic times, companies will invest in technologies that merely "have a chance" to succeed. For example, many venture capital firms funded Internet startups with any hope, and in bad economic times curbed funding because their propensity for risk changed. A compact way of looking at the evolution of IT investment ideas in most businesses is contained in the following chart:

Evolution of Ideas in Business

Breakaway Opportunity *Pioneer*	Competitive Opportunity *Early Adapter*	Competitive Necessity *Mainstream*	Operational Necessity *Cookbook*
- High Risk - Long time to implement - *Potential* ROI	- Moderate Risk - Moderate time to implement - *Proven* ROI	- Low Risk - Short time to implement - *Moderate* ROI	- No Risk - Implement for survival - *Negative* ROI if not done
Example - PC's in 1985	Example - PC's in 1990	Example - PC's in 1995	Example - PC's in 2000

Time →

Note that the risk is highest and the advantage is greatest for the pioneers and that there are more pioneers willing to take risks in good economic times. Another way of looking at this continuum is that over time what was a pioneering technology gradually becomes mainstream and cookbook. The evolution of Internet utilization is depicted in the following exhibit:

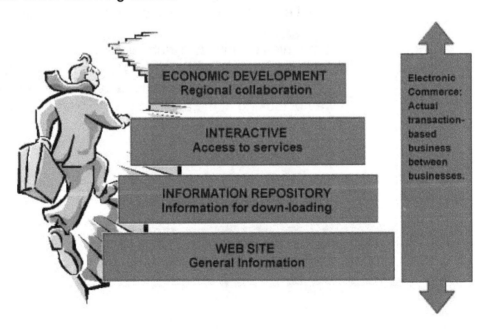

According to Gartner's Research Note entitled "Selling IT Investments to Business Leaders" (DF-14-3337, October 10, 2001), IT can choose five basic justifications when positioning an IT investment project:

Business Justification of IT Projects

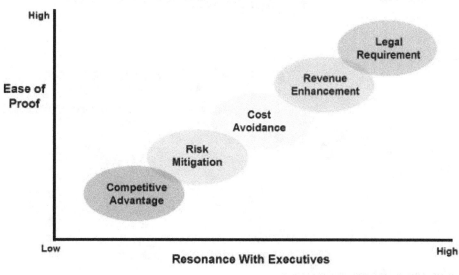

The closer the project is to a legal requirement, the easier the justification and the greater the resonance with executives.

In any case, Drivers create the initial organizational "shock waves" that in turn create statements of goals and objectives in Strategic Business Plans that exist in every organization.

Chapter Two

Business Strategy

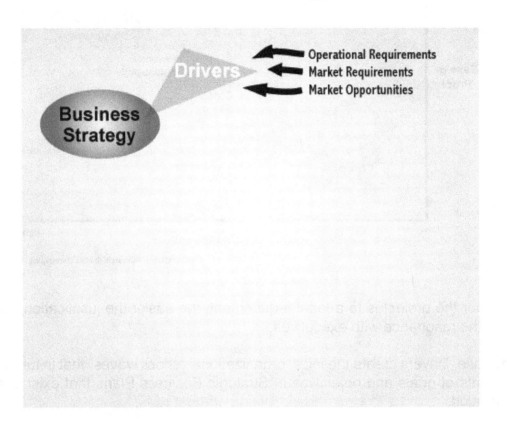

Ask Yourself:

Does my company have a strategic business plan? Is it shared with as many employees as possible?

In Gartner's March 2001 issue of EXP Premier "Leading Through Tough Times, " the top ten CIO issues are:

1. Strategizing for IT/Business Linkage
2. Attracting and Retaining People
3. Leadership and guidance for board/executives
4. Nurturing and sustaining IT competencies/people
5. Demonstrating business value of IT
6. Developing e-enabling IT architectures
7. Program/project management
8. Prioritizing projects
9. Developing the IT organization as a professional services business
10. E-business process management capabilities

The following trends didn't make the top 10 but at another socio-economic time they might:

- Globalization
- Greater emphasis on people and life balance
- Growing power of brands
- Market deregulation/regulation
- Increasing number of mergers/acquisitions
- Breaking down barriers between markets
- Rapid scaling of prototype business ideas into big business
- "Dis-intermediation," (creating new techniques to remove specialized brokerage expertise requirements from a customer/vendor relationship)
- Competitors forming alliances
- Move from proprietary to open trading networks
- Forming alliances with competitors
- Flexibly moving into and out of markets
- Growing litigation (Shareholder, customer, or alliance partner)

Ask Yourself:
Which of these issues are business issues and which are IT issues?

I score them this way:

- Strategizing for IT/Business Linkage (IT)
- Attracting and retaining People (IT)
- Leadership and guidance for board/executives (Business)
- Nurturing and sustaining IT competencies/people (IT)
- Demonstrating business value of IT (IT)
- Developing e-enabling IT architectures (IT)
- Program/projects management (IT)
- Prioritizing projects (Business)
- Developing the IT organization as a professional services business (IT)
- E-business process management capabilities (IT)
- Globalization (Business)
- Greater emphasis on people and life balance (Business)
- Growing power of brands (Business)
- Market deregulation/regulation (Business)
- Increasing number of mergers/acquisitions (Business)
- Breaking down barriers between markets (Business)
- Rapid scaling of prototype business ideas into big business (Business)
- "Dis-intermediation," removing intermediaries (Business)
- Competitors forming alliances (Business)
- Move from proprietary to open trading networks (IT)

9

- Forming alliances with competitors (Business)
- Flexibly moving into and out of markets (Business)
- Growing litigation (shareholder, customer, or alliance partner) (Business)

For heads of IT departments, the IT issues generally seem to rank higher than business issues. This is the opposite of the way it should be. In contrast, in "The Chief Executive Guide to the Internet," a supplement to Chief Executive Magazine, "Five Questions to Ask your CIO" appeared regarding investments in the Internet:

1. What specific new information will we receive from this investment?
2. How will this investment make my employees better at their jobs?
3. Which division is driving the expenditure?
4. Who will be effected by the change and how long will they be drawn away from their primary jobs?
5. How will the new system build on our present organizational infrastructure?

Note how all of these questions are related to the business and not to technology.

As Yourself:
What is the relationship of the CIO to the other chief executives of the organization?

A partnership needs to be established between the head of the firm's IT group, normally titled the CIO (Chief Information Officer) or CTO (Chief Technology Officer), and the heads of finance, marketing, operations, and legal affairs. As part of the executive team, the CIO/CTO can input technology capabilities to the business strategies but should never make technology an end-result or goal of the plan. Business Strategic Plans are best built when the partnership between IT and the other chief executives is informal but real. If the CIO/CTO reports to the CFO, for example, the relationship with the others can be diminished.

Ask Yourself:
Does my company segment the business plan is an easy to view format as the exhibited below:

GOALS→STRATEGIES →TACTICS
Few → Several → Many
General → More Specific → Most Specific

A good way to look at Strategic Business Plans is that a firm should have a few, clear goals, several strategies to accomplish those goals, and many tactics to achieve the strategies. The transportation company I worked for expressed its business plan with five high level goals, 39 strategies to achieve the goals, and 165 tactics to achieve the strategies. The IT elements of the plan were embedded tactics to support strategies. Technology was never expressed as a goal. In manufacturing, my experience was the same.

My banking experience was different. To start-up a commercial loan business, the technology infrastructure was a pre-requisite and therefore a tangible goal for without its achievement, loans could not be originated, underwritten, or serviced. Once the infrastructure was established and the commercial loan business was firmly underway, the IT element of the business plan took its traditional place - sub-serviant to business functions. As Bruce Rogow states in Gartner's Marketspace Odyssey 2001, "The challenge for CIOs is to restructure their agendas away from implementing one mega-project to looking for the points of leverage among several thousand smaller projects." Consider the matrix below; notice how goals and strategies can be the same across industries even if systems are different:

		Banking	Manufacturing	Transportation	Government
Goals	1	Be the best price in the market	Improve quality to best in class	Improve Service	Break-even
	2	Be low-cost provider	Be low-cost producer	Increase ridership	Treat citizens as customers
	3	Be best in customer satisfaction	Expand markets	Enhance quality	Increase job-growth rate.
Strategies	1	Improve systems	Improve systems	Improve systems	Improve systems
	2	Enhance staff training	Enhance staff training	Enhance staff training	Enhance staff training
	3	Improve customer communications	Improve customer communications	Improve customer communications	Improve customer communications
Tactics	1	Create integrated suite of loan processing systems including: originations, underwriting, & servicing	Create integrated suite of process control systems including: customer/order management, just-in-time-inventory, & raw materials scheduling	Create integrated suite of route optimization systems including: scheduling, vehicle maintenance, & ridership accounting	Create integrated suite of citizen awareness systems including: tax filing, voter registration, & license management
	2	Create training database and needs analysis for each employee	Create training database and needs analysis for each employee	Create training database and needs analysis for each employee	Create training database and needs analysis for each employee

In summary, a strategic business plan of one-, five- or ten-year terms arises from drivers. Embedded in these strategic business plans are tactics to achieve success. Within these tactics are the roots of the **IT Application Portfolio Plan and the Technology Triad.**

Chapter Three

Technology
Planning Triad

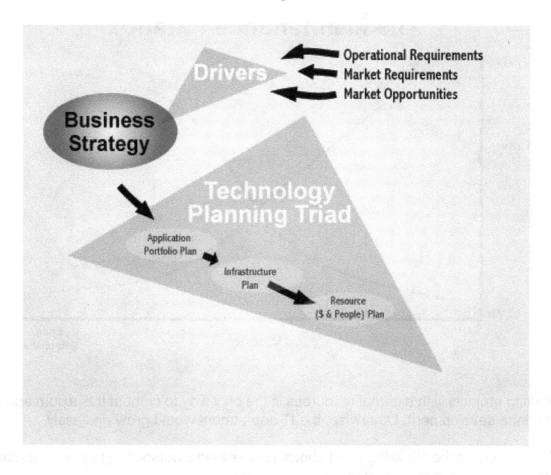

A s systems are moved to completed status, the amount of staff time available for new systems decreases as the graph below suggests:

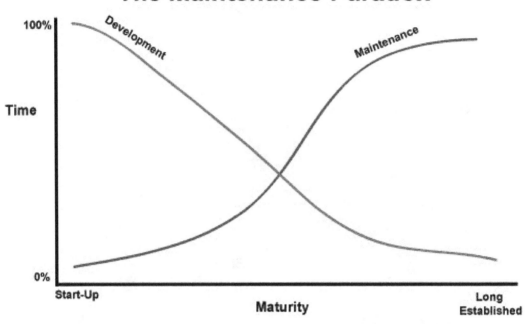

The Maintenance Paradox

Staffing projects with external resources is the only way to combat this axiom and to continue development. Otherwise, the IT department would grow endlessly.

The overlay on the following chart shows how strategic outsourcing of any non-core functions can cure the paradox.

Solution to the Paradox...
Strategic Sourcing

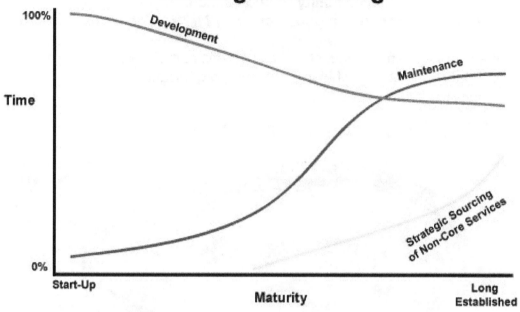

Ask Yourself:

Is my company ready to support IT maintenance requirements? How will this cost be charged to the consumers? How will this effect the company's ability to be flexible in a bad economy or to support innovation in a good economy?

As the following exhibit suggest, IT has become part of the 21ˢᵗ Century Business Model and an <u>expected</u> part of high-performing organizations

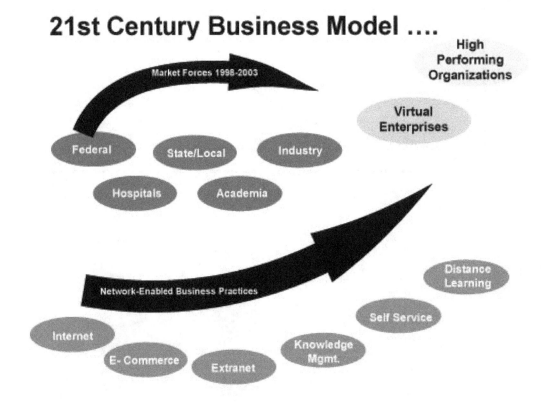

To integrate IT with the Business Strategies of an organization, the IT Triad needs to be undertaken. The IT Triad is made up of the Application Portfolio Plan, the Infrastructure Plan, and the Resource Plan. Each element of this triad requires sponsorship at the highest levels of a company:

- application portfolios require blessing of priorities sometimes making difficult choices between competing strategic business units;
- infrastructure plans require executives to understand the current and future ramifications of picking certain technologies and technology companies as partners; and
- resource plans require management to budget for the IT expenditures integral to the success of project and in turn integral to the fruition of the business value added by the project.

Each element of the triad receives a subsequent chapter in this book.

Chapter Four

Technology Planning Triad:
Application Portfolio Plan

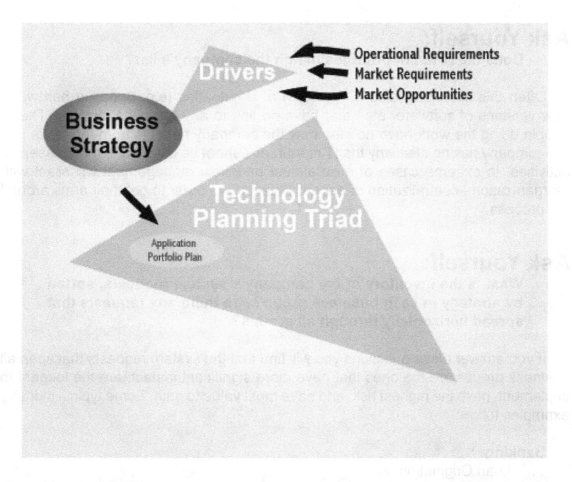

In 1984, the American Productivity Institute conducted a study to see if Americans are different than Germans or Japanese in their approach to work. Their experiment consisted of a model airplane kit being handed to a child of each of those cultures. The German and Japanese children tried to "do it right the first time" by opening the box properly, reading the instructions, and proceeding to completion of the model sequentially from first step to last. The American children created a different impression. They had a tendency to rip open the box, barely glancing at the instructions, and building their "best guess" of what the model should look like often performing steps out of sequence. Noteworthy was that their speed to completion was significantly faster than that of the other two cultures.

A conclusion from this experiment is that Americans do not have a "do it right the first time" attitude towards work. Rather, a continuous improvement attitude is more likely to succeed when Americans are involved.

Application Portfolios are merely inventories of systems needed by an organization in support of business initiatives. The systems that are of the continuous

17

improvement nature (usually the ones that provide operational requirements) are readily discernable and prioritized. The systems that create new markets or drastically alter the nature of some aspect of the business are not as easily focused. The essential question to pose is:

Ask Yourself:
Does the IT list of priorities match the company's list?

Often this exercise yields a mismatch of IT priorities (e.g. updating hardware, new versions of software, etc.) that have no link to any business strategy. The IT people doing the work have no idea how the company benefits from the tasks and the company has no idea why the IT resources cannot be diverted to more essential activities. In extreme cases of mismatched priorities, management will react with reorganization - centralization or decentralization - in order to get their arms around the process.

Ask Yourself:
What is the inventory of my company's system requests, sorted by strategy in each business group? Are there any requests that spread horizontally through all groups?

If you answer these questions you will find that the system requests that span all business groups are the ones that have more significant impact, are the longest to implement, have the highest risk, and have most value to gain. Some typical industry examples follow:

Banking:
Loan Origination
Underwriting
Servicing
Pooling

Manufacturing:
Customer/Order Management
Raw Materials Scheduling
Activity-based costing
Shipping/Receiving/Invoicing

Transportation:
Route Scheduling
Vehicle Maintenance
Customer Travel Advisory
Fund Accounting

Ask Yourself:

What are the elements of successful multi-project, application portfolio management?

- **Maintain a master plan by strategic business group** - simply reorganizing the full portfolio plan into strategic business sub-lists is not the entire task. Managing expectations of functionality to be delivered, timing of delivery, performance levels to be achieved and return on investment are at the top of most executives wish lists. These have to be mapped into the plan so that the resource level of effort is known by skill set and source and so milestones can be established with some certainty. Reviewing the sub-list with members of the strategic business group on a periodic basis is the ultimate answer. These meetings ensure that the IT Department and the business folks are on the same page, literally and figuratively.

- **Consistent project methodology** - there are many methodologies a firm can purchase and adopt. Eventually, a company's project methodology evolves into one that is unique to their internal processes. For example, government procurement processes have been shaped by decades of regulatory enhancements (such as those made for Disadvantaged Business Entrepreneurs) that need to be recognized within a project timeline. On the other hand, all methodologies have elements that are common. They all progress through a continuum of

Plan□Design□Build□Implement

These PHASES all have TASKS, which in turn have STEPS, and at the end of each PHASE, approvals are required. General Electric has elevated this to a cultural imperative called "Six Sigma," and it is ingrained into every project, IT or otherwise, undertaken in each part of the firm. To be a "Black Belt" (the highest certification in quality) in Six Sigma is an achievement, and to be unfamiliar with quality project techniques and methodology is to be employed elsewhere. A more general methodology, which I adopted in manufacturing, finance, and government, appears in the following exhibit:

System Development Methodology

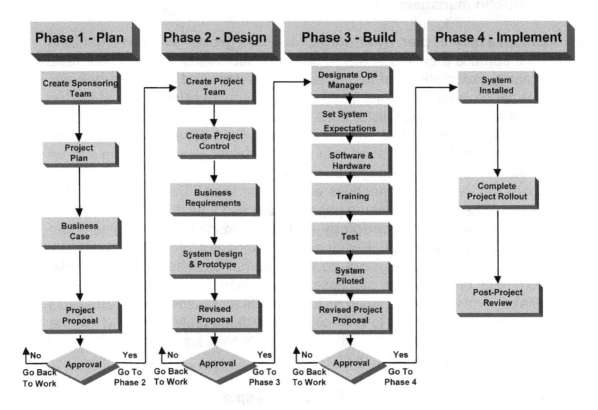

- **Tracking achievements against expectations** - internal customer expectations and perceptions of how the project is going is most reflective of how IT matches the company's needs with the project portfolio. There is a possibility that projects are led astray by business sponsors who interpret the company's needs differently or differently through time. However, there is a lower risk of that happening than misinterpretation by the IT Department. Clearly, top management can be in agreement but execution can be distorted by the most well-meaning managers. That is why there must be a constant alignment of expectations with targeted performance for each phase of the project in each dimension - functions delivered, timing, customer's rating of performance, and cost.
- **Tracking expenditures against budget -** every business tracks expense versus budget because it is prudent to do so. Nothing embarrasses more people in an organization than a project that has spiraled out of control financially. Control by phase (plan, design, build, implement) is best because it allows tuning over the lifespan of the project instead of risking being out of funds at crucial implementation junctures.

- **Tracking staff available and staff augmentation methods** - The longer the project, the higher the risk associated with the human resources assigned to the project. Beyond a six-month period, staff stability becomes a very real issue. To safeguard against personnel transitions, project managers need to have a ready supply of staffers or outside sources at all levels to fill in any gaps. Thus, a blanket staff augmentation contract, whereby a firm creates an overall agreement to procure temporary personnel at rates associated with their position and then executes task orders within the agreement, should be one of the initial items of the Technology Planning activities. Also, development activities need to be coordinated with the HR Department and other staff to make sure that other departments are not planning resource changes or new employee programs designed to allow career management at the employee level that will conflict with business projects.
- **Providing deliverables along the way** - A simple but effective method of keeping executive management satisfied that a project is moving along at an expected pace, is the delivery of some feature of the system prior to the entire completion of the system. For example, a data network for entire firm of, say 10,000 employees could take years to fulfill in its entirety. By implementing whole divisions or departments along the way with new desktop PCs as their turn comes up in the full project plan is a good way to win friends to ward off enemies. Always making progress is a good way to keep company perceptions positive. These progressions have to correspond to useful deliveries but that usually is not hard to find because processes addressed by these are usually poor to begin with.
- **Coordination of hardware, software, and network tools**
- **Understand interdependencies and interfaces**
- **Guard against scope "creep"**
- **Ensure that business representatives and project teams fully communicate regularly**

Chapter Five

Technology Planning Triad:
Infrastructure Plan

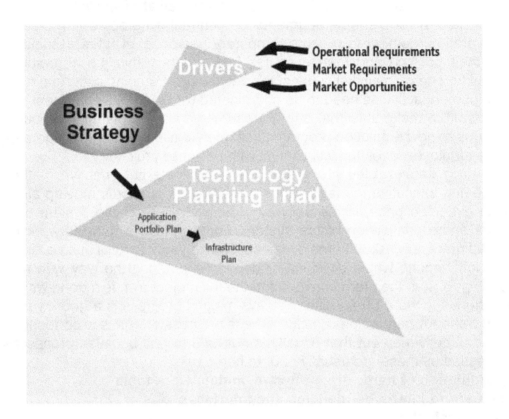

The infrastructure planning phase is, unfortunately, where many technology leaders like to start. Their knowledge of hardware, software, database, and network components screams to get out of their heads and into the ears of executives with funding authority. It is a temptation for many CIOs and CTOs to sing the praises of the latest chip speed, or the newest network router. In truth, it is the Application Portfolio Plan that will dictate the *Infrastructure Plan.* There will be very little choice - the software or the software company will dictate the hardware.

Ask Yourself:
Can we find software that matches our requirements?

In my Manufacturing Industry experience, we were looking for a system that could integrate several applications - customer-order processing, raw materials scheduling, activity-based costing, just-in-inventory management, and supply chain management. We reviewed the products of several software companies, each running on a different hardware, network, and database platform. When we made the final software selection, the infrastructure the vendor recommended (usually the one on which they created and developed the software) was the one we selected.

This happened to be an IBM-centric platform. Although technical arguments could be made for DEC and HP platforms at that time, selecting the other platform vendors would add risk to the project for no reason other than "the technology people would prefer something else." The business case for selection of the software negates any technical case for the infrastructure in almost every occasion.

Ask Yourself:

If we can't find a single software product matching our full requirements, can we find common infrastructure components within several software products?

In my banking experience, we were looking for a system that could do loan origination, underwriting, servicing and securitization. When no single software could be found to do all components, we looked for several "best of breeds" that could interface with each other. We selected a combination of vendors that each used a Microsoft database product - so that at least the data was on one platform. While IBM had the best loan servicing system it was on their proprietary database. So the firm selected the next best COMPATIBLE software for the loan origination and underwriting systems.

Ask Yourself:

Aren't there some occasions when infrastructure support limits the possible number of software selections?

In a transportation project, we limited our software search based on the business requirements first (of course) and then on the fact that the software needed to use diskless PCs on the depot floors for ease of maintenance. This limitation created a shorter list of software vendors from which to select. When we found the right software, the infrastructure components (in this case, IBM, Oracle, and Nortel) were dictated.

There are many technologies available in the marketplace - dozens of PC manufacturers, thousands of software companies, scores of network providers. Selecting a technology is very difficult if that's what one sets out to do. On the other hand, if selection starts with the business requirements, and then software is found that matches those requirements, hardware, database, and network infrastructure platforms EVOLVE. In another sense, the infrastructure "selection process" is an oxymoron in that any that the software vendor dictates can work and in that there really isn't a process.

Ask Yourself:
What if we have a general platform and new software is derived that dictates a different platform?

Difficulties arise when several projects derive the same platform (de facto standard) and a stray project derives a different platform. CIOs earn their money in these situations by either forcing the software selection group to pick a second-best product that meets the de facto standard or by getting the resources necessary to support the stray selection.

Ask Yourself:
Why are standards important?

Standards are important to leverage resources, both capital and human, optimally. If companies select "one of each" or set up a systems administration for each and every application in the portfolio, the amount of money and associated labor would send the minds of executive management and board of directors spinning. The concept of systems' optimization reminds me of a story I heard by Emory Zimmers, PhD., head of Lehigh University's Computer-integrated Enterprise School:

> The Professor asked his Freshman Engineering students:
> "If you have a five step process, and you optimized each step to its fullest, have you optimized the process to its fullest?"
> All of the Engineering students said yes. A basketball player auditing the class was the lone dissenter. The professor asked the athlete: "Why do you disagree with the rest of the class?"
> The athlete replied: "On a basketball team, if we had the very best player at each position, we wouldn't necessarily have the best team, in fact, we might have chaos. For example, each player would want the ball in order to score and we wouldn't play team defenses like zones very well either. Sometimes, we need players to play in a team concept and subjugate their egos. Likewise, I think that an individual process must be sub-optimized in order to optimize the entire process."
> The professor smiled and invited the student to make engineering his major course of study.

A typical top ten view of IT priorities as it applies to Infrastructure Planning can be found in the Gartner Publication "Leading Through Tough Times - CIO Agendas 2001" as follows:

1. Internal e-business architecture
2. Enhancing security
3. Network architecture management
4. Building inter-enterprise e-business processes

5. Application integration
6. Implementing customer relationship management
7. Applications scalability
8. Deploying enterprise portals
9. Desktop management
10. Windows 2000 deployment

In the list above, there is a clear reference to the Technology Infrastructure concerns IT executives must endure.

In the same publication, IT management priorities appear as well with less emphasis on technology and more emphasis on business concerns:

> **Strategizing** for IT/Business linkage
> Attracting and **retaining** people
> **Leadership** guidance for Board and executives
> **Nurturing** IT competencies
> Demonstrate the **business value** of IT
> Developing **e-enabling** architectures
> **Project management**
> **Prioritizing** projects
> Develop IT organization as a professional **service** group
> E-business **process** management capabilities

Once the application portfolio and infrastructure plans are establish, IT management can begin to put a price tag on the IT overall plan. This **Resource Plan** needs enterprise-wide collaboration.

Chapter Six

Technology Planning Triad:
Resource Plan

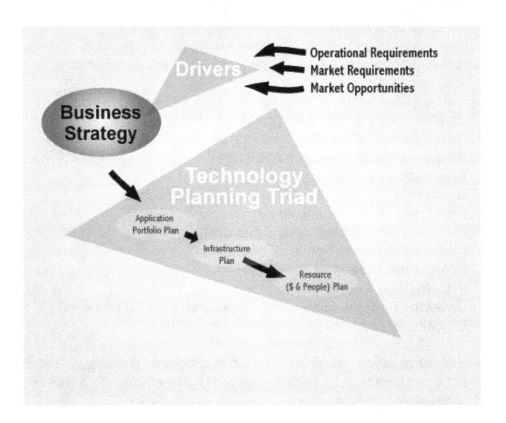

Getting the resources necessary to start and complete IT projects has always received a great deal of scrutiny in all industries. In the early days of computing, justifying IT projects was relatively simple and Return on Investments (ROI) was couched in reduction in labor or more accurate record keeping. Today, going from one automated system to another, albeit newer, system is not as easily justified.

New-age justifications lie within better work processes which seem much more intangible to the questioning Chief Financial Officers. The lack of problems surrounding the century date change caused many a cost accountant to sneer cynically and one can almost hear them say ". . . we spent all that money for nothing." As a result of the "Y2K" non-events, the scrutiny of IT projects has become exponential. One CIO I know told me that "it is not the return of the bean counters; it is the revenge of the bean counters!"

With all of this suspicion and haranguing over IT project development costs, few ask the question "What happens if we do nothing?" Clearly, some businesses might die if they stay with old technology.

However, if the IT Planners have created a foundation for asking for resources, the resources do become available even in the most skeptical companies. The foundation is created by:

- Certification and acknowledgment of business drivers
- Collaboration in the creation of the strategic business plan
- Completion and prioritization of the application inventory
- Knowledge of the infrastructure requirements and standards that satisfy the highest application

Ask Yourself:
How do I get the people I need to fulfill the application inventory?

The people equation is managed in only a few ways. Staffing for a project is allocated from current headcount, is newly recruited into the company, or is hired on a temporary basis. Naturally, the final result is usually a blend of all three scenarios. Project teams consist of these blends both from an IT prospective and from a business group perspective. Internal staffs members are the toughest to control in terms of time allocation especially as maintenance of current systems remain their top priority.

An example of how a business strategy flows into the Technology Planning Triad, which in turns leads to the budgeting of resources, is depicted below:

Strategy — Improve customer communications

Goals — Promote excellence / Make the company the vendor of choice

Application Portfolio — Improve Website / Integrate Financial Suite / Improve Telephone information Center

Infrastructure Plan — Improve Data Network

Resource Plan — Capital Funding - By Project, total $40 million / Operating Funds - By Project, total $20 million / Personnel - 40 transfers, contractors, new hires / Training - Technical & Managerial, $1 million

Ask Yourself:

How do I maintain systems and initiate new systems development at the same time?

The answer to this question involves a complex series of relationships but boils down to one thing - always ensure that there is someone to call if the system breaks. The structure that has worked for me is depicted as follows:

System Leader
    → **System Analyst**
        → **System Programmer**
        → **System Vendor support**

Of course, these positions vary by size & scope of system and they do not take into account 24 x 7 data center groups who do general maintenance like backup copying and batch printing. The key ingredient is the vendor support, which is easily flexed to include all support encompassing analysis, programming, trouble calls, and data issues or to simply provide a hot line to call in an emergency. The point is that:

NO SYSTEM CAN REMAIN WITHOUT SUPPORT AS LONG AS THERE ARE USERS WHO MAY NEED HELP.

In manufacturing, we had an order entry system that operated in three time zones. Staffing was added to accommodate. In Banking, we built systems with no staffs but full vendor support, which mandated very detailed service level agreements. In Transportation, we had a blend that became more vendor-support oriented over time. The following chart portrays the typical structure of personnel within an IT Department:

Project Services
- **Director → Project Leader → Systems Analysts → Programmers**
  → Vendor Support Contracts

Data Center Services
- **Director → Shift Supervisors → Operators → Helpdesk**
  → Security Manager → Disaster Recovery Manager
  → Operating Systems Specialists
  → Vendor Support Contracts

Network Services
- **Director → Team Leaders → Network Technicians (Data & Voice)**
  → Vendor Support Contracts

Ask Yourself:
How do I get the budget necessary to fulfill the project plans?

Money is often more easily obtainable than personnel. Once a business driver generates a business strategy, all ideas (with or without IT content) are promoted to achieve the desired results. In Manufacturing, when a new plant was approved in order to compete with the market leader, the virtual reality software used to visualize the plant and simulate the processes was just a $250,000 afterthought. In banking, once a decision was made to handle loan escrows for tax and insurance for up to 10,000 loans $30 million and higher each, the $1.5 million price tag for a loan servicing system was a necessity not a luxury. In Transportation, the $2.5 million cost of a vehicle maintenance information system was offset by the promise of more buses being in service rather than sitting in service bays. At $300,000 per bus, only nine buses had to be placed back into to service to cover the cost.

Chapter Seven

Technology Planning: Rapid Deployment Tactics

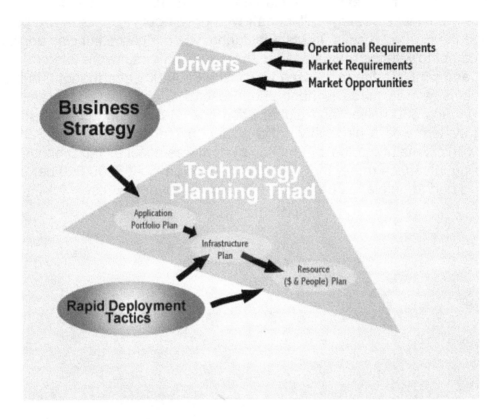

There are many ways to increase the speed at which projects are completed. A former boss of mine once said to me that most projects have less than a year of "real life" defined as the time where they are considered successful or not. Time is never on the side of a project - budgets change, executive sponsors come and go, team members are offered better opportunities, competitors develop new systems that render your project superfluous, or the company consciousness is flooded with higher matters.

In my experience, the development of the Application Portfolio Plan has its own timeframe and it will get done when it gets done. It is a function of the number of interviews to be undertaken and the number of current systems needed to be assessed. However, there are dozens of ways to speed infrastructure and Resources Plans for individual projects as well as all the company projects collectively.

Ask Yourself:

How do I speed the delivery of my projects? How does the company view speed as a factor in the company's fortunes?

There are some generic speed tactics within IT circles as depicted in the following list:

Strategic Sourcing & Rapid Deployment Techniques

- **Off-The-Shelf Software**
  –Development with external Application Software Providers
- **Fully Loaded Hardware Purchases**
  –Include Set-Up, Migration, Maintenance, & Disposal
- **Internship Programs**
  –Staff Augmentation In Exchange For Workplace Experiencce
- **General IT Consultant Blanket Contract**
  –On-Demand Staff Augmentation Without Long-term Commitment
- **Managed Services**
  –e.g. Web Site Hosting, Paycheck printing/distribution

In today's IT environment, Application Service Providers (ASPs) have been used to ramp-up new applications quicker than an internal staff could. The new age twist to this outsourcing technique is that the internet allows companies to log on to application from day one of the project instead of waiting for the traditional set up of hardware and software. In this manner, edits and/or customization of the software to fit the company can begin right away. Examples of what may be obtained from ASPs for schools, airports, and transit follow:

Application System Providers: Airports

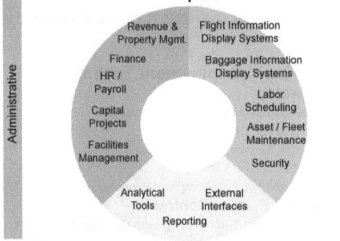

Application System Providers: Transit

There are simpler methods of increasing speed. I have used several, additional speed tactics that seem to work almost every time:

"Produce Deliverables along the way"
- If a project is scheduled to last one year, management senses urgency within a short time horizon. If the project team can deliver something, maybe anything, to the project sponsors before the one-year mark, managers feel reassured that progress is being made.

- Example in Manufacturing - While we were busy planning, programming, and creating an enterprise-wide financial suite, within the first six months of the project we discovered duplicate payments within the existing Accounts Payable system. A simple report from the data already existing in the system uncovered over $250,000 in duplicates. While the CFO was a sponsor of the new project, the discovery allowed him to say, "I told you so" and re-enlisted the rest of management to support the new project.
- Example in Banking - While developing a notebook PC application for traveling loan officers, we first taught them how to get on-line. The training alone made them more efficient, and fostered their allegiance to the new upcoming applications.
- Example in Government - When installing new digital copiers with network printing, scanning, and faxing features, we offered the secretaries the features one at a time in order to build up their confidence in the new "high-tech" equipment.

Avoid "Administrivia"

- There are many ways for the "old way of doing things" to block the path for the new way or to at least detour a project. Unwittingly or sometimes maliciously, company staffers will create these roadblocks. They can come in many forms - meetings about the old system when everyone knows a new system is coming in; enhancement requests for the old system when there should be a freeze on enhancements; or withholding of signatures due to an inability to understand the reason for the system. Many of these roadblocks come from the staffers who have a vested interest in keeping things the way they are and have no enthusiasm for the new project or system it produces. In these cases, project managers and their sponsors need to keep focused on milestones and forge ahead. Make no mistake, some people want to see the project fail.
- Example in Manufacturing - While we were getting close to completing the financial suite for one strategic business unit, another SBU requested more time to investigate the nuances of the new system. In truth, they had a home grown system in which many staffers had pride in authorship. From a business standpoint, the home grown system could not compare with the features of the commercially available system but the delay cost the project several months.
- Example in Banking - A project that intended to provide notebook PCs for loan officers was successful unto itself but interfacing the notebooks to the back office systems proved problematic. The outsourcing company who "owned" the back office system, complained they had other priorities above the notebook interface. In truth, they saw the notebook system as competition to their owned desktop services. So much time was lost, that the project phased off the corporate radar screen when economic times changed so that the notebooks were not available for all officers.
- Example in Transportation - When installing a new vehicle maintenance system at 12 bus depots, one depot had repeated incidents involving theft

of computer equipment and sliced wiring. The depot's staff feared that the system would track maintenance work against standards and make workers accountable for various issues. In the end, the project was completed only slightly behind schedule and the mechanics' staff was reduced by 10%.

Use appropriate Humor and/or stories to relieve tension and to drive home lessons to be learned

I have found that the leader of a project or of a department must be cool under even the most stressful of situations or deadlines. One method of demonstrating coolness is with anecdotal humor, using (but not overusing and certainly not inappropriate) stories and quotes to convey a message to the team. Some of my favorites follow:

- Stories
- **Casey Stengel & Left Field:**

Casey Stengel was a baseball player for a long time. Before he became a full time manager, he spent a short time as a player-coach spending most of his time in the dugout but sometimes placing himself in the field. One day, a rookie left fielder was having a terrible time fielding the ball and in fact made three errors during the game. Before the last inning, Casey took the kid out and placed himself in left. As luck would have it, a ball was hit to Casey and he promptly misfielded it for another error. When he came back to the dugout at the end of the inning, the smiling rookie was staring at him. So Casey said: "Kid, you've got left field so messed up, nobody can play it."

Lesson - Don't be so quick to criticize.

- **The Cardinal and the broken microphone**

A Cardinal was preparing for a mass at the main Cathedral in Philadelphia. There were about 1,000 faithful in attendance who knew how to respond at mass. For example, at the end of a mass, the priest will spread his arms and say "Peace be with you" at which the assembly responds "And also with you."

On this day, the Cardinal had a troublesome microphone attached to his robe. Sometimes it would work fine and other times it would simply turn off. All mass long, the Cardinal was either screaming to be heard or speaking normally with amplification from the microphone. By the end of the mass, however, the microphone gave out altogether and did not come back on. So the Cardinal spread out his arms wide and said "There's something wrong with this mike" at which the assembly responded "And also with you."

Lesson - Make sure instructions are heard clearly.

- Witticisms
- Mark Twain:

"All generalities are false, including this one."

Lesson - Do not be prone to grand, general statements (like the system doesn't work). Be specific when uttering complaints.

- Yogi Berra
  "It ain't over 'til it's over"

  Lesson - All projects need closure but projects are closed only when the user says they are closed.

  "It's déjà vu all over again"
  Lesson - We can learn from project experiences of others no matter how different the project.

  "90% of this game is half mental"
  Lesson - Projects don't fail because of the technology, they fail because of people.

Manage meeting time effectively
The greatest wasters of time for projects are meeting especially the ones that have overflow crowds.

Ask Yourself:
Are your meetings wearing everybody out? Of the time you invest in holding meetings, how much actually translate into profitable results?

The fact is more than 50 percent of the time spent in most meetings is completely wasted. This is true regardless of the type or size of your organization. How good are your meetings? The following strategies, passed to me by Paul Schmitz of the Gartner Group, will help you get the most out of your team sessions:

10 steps to more effective meetings

1. <u>Be ready.</u> **Think through the purpose and objectives for the meeting. Be crystal clear on what you want to get out of the meeting before it convenes and how you plan to get there.**
2. <u>Tell them, too.</u> **Prepare people who will be there. Explain the purpose, agenda and what you expect of everyone in the meeting. Do this so they can prepare, too.**
3. <u>Get the right people in the meeting.</u> **You need three kinds of people in every meeting: those with authority, those with expertise and those who will be "doers."**

4. <u>Don't forget the logistics.</u> Flip charts or white boards serve as the " common brain" for the meeting. If you don't have a common brain, everyone will use his or her own. Lighting, noise control, comfortable seating and the sides of the room are also important to consider. In addition, remember that task-oriented meetings that go on for more than an hour lose their effectiveness.

5. <u>Set ground rules.</u> Establish rules ahead of time. Good rules include starting and ending on agreed times, minimizing interruptions and allowing no one to dominate the meeting.

6. <u>Take minutes.</u> Put at least three flip charts on the wall before the meeting. One is for the agenda, one is for issues that need to be discussed later, and one is to record decisions and action items made during the meeting.

7. <u>Wrap-up.</u> Make sure your reserve enough time to summarize the results, decisions and assignments. The three charts will come in handy for this purpose.

8. <u>Publish minutes immediately.</u> They don't have to be detailed simply include the highlights of the major decisions and action items and get them out quickly (within 24 hours).

9. <u>Follow-up on assignments.</u> Hold people accountable for the results. Follow-up will also tell you if what was decided it was or is a good idea or not.

10. <u>"Tune-up" your meetings.</u> Always strive for improvement. Meetings are often doomed to mediocrity unless people work on them. Have the participants write down what they don't like about the meetings. Then list their complaints on a flip chart before you begin discussing them. Flip charting the issues gives everyone the overall perspective of what needs to be worked on.

Chapter Eight

Technology Planning: Value Added Systems

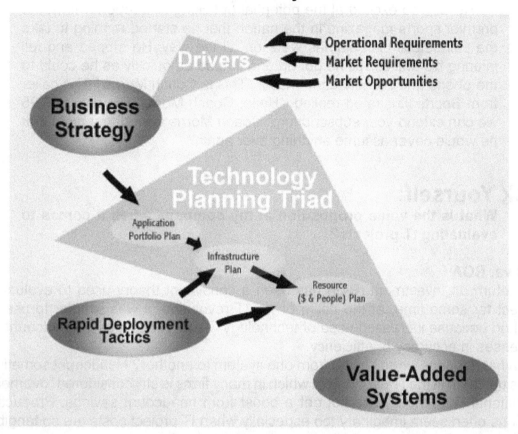

Ask Yourself:

Is IT guilty of creating false expectations and assumptions?

False expectations or the syndrome of "over-committing and under-delivering" can spell trouble for heads of IT Departments. I often retell as story told by former LaSalle University's Men's Basketball Coach Speedy Morris that describes false expectations:

Story—Speedy Morris was named head coach of LaSalle's Men's Basketball team - a job he dreamed about his entire life. On the day he was named coach, Speedy was interviewed by all three of Philadelphia's major TV stations and by the city's two leading newspapers.

After a full day of media attention, Coach Morris returned home and announced to his wife that he was going to bed early because of his exhausting day. While he was putting on his pajamas, the telephone rang.

She answered the call and announced "Speedy, Sports Illustrated is on the line and they want to talk to you."

Coach Morris thought to himself, "Wow! Sports Illustrated - I must have really hit the big time!"

He was so excited at the potential of talking to a reporter form the premier sports magazine in the nation that he started running to take the call before his pajamas were on all the way. He tripped and fell injuring his ankle. Still he got up and limped as quickly as he could to the phone and proudly announced, "This is Coach Morris." The caller from Sports Illustrated replied, "Hello, Coach Morris. For only $29.95 we can extend your subscription." Coach Morris vowed that night that he would never assume anything ever again.

Ask Yourself:
What is the value proposition at my company when it comes to evaluating IT projects?

ROI vs. ROA

Return on Investment (ROI) has been a consistent theory used to evaluate IT project for some time. At the dawn of the IT revolution, it was simpler to use this method because increased used of technology meant reductions in headcount and increases in accuracy or efficiency.

What happens when you go from one system to another? Headcount sometimes goes up, often in the IT department which in many firms is still considered "overhead." Traditional ROI, thus, does not get a boost from headcount savings. Productivity savings often seem imaginary too especially when IT project costs are so tangible - hardware, software, network, license agreements, maintenance agreements, etc. It is relative easy to determine the cost of the new system over time but very difficult to quantify the savings using traditional methods.

Some practitioners feel that Return on Assets may be a better barometer. In this method, the value of the a new project or system comes from the way it extends current productivity whether it be increased labor efficiency, enhanced utilization of real estate/plants/office space, or improved processes as in supply chain management or customer relationship management. This method creates a new way of thinking about technology as it applies to the business but it is really the extra seal of approval that intuitive project sponsors seek when they announce their desires for the project. Asset Management Systems are invaluable for forecasting which project will benefit from ROA-style justifications.

Making judicious IT investments clearly allows firms to move forward - IT is a natural catalyst for change. Even when a company wants to "stand pat," IT can help reduce costs. Early IT/computerization projects usually had reduced headcount as a clear factor in ROI calculations. Today's systems have less visible value. To that end, ROA (Return on Assets), i.e. getting more value from existing assets, may be a better measuring tool.

Example of a value-added system . . . vehicle maintenance information.

As expressed by an external consultant, a transportation company had significant operational issues that warranted attention:

- vehicle out-of-service rates were as high as 33%, compared to single digit rates in industry "best practices" companies
- no work standards per work order type had been established in any of 12 depots for any mechanics resulting in lost hours of effort and other inefficiencies;
- warranty tracking needed improvement for expensive new bus fleets which cost about $300,000 per bus; and
- an abundance of mechanic downtime was due to "awaiting parts" because the old system had no interface to the inventory and purchasing systems.

We decided to implement the best "off-the-shelf" vehicle management information system available in the Transportation Industry, in our opinion. The procurement process (including RFQ, RFP, negotiations, and contract approval) was completed within only three months and included collaboration with significant managers within the organization including the COO and CEO.

We installed the prototype at one depot within 8 months. The solution was then rolled-out to the 11 remaining depots within 12 months. This system contained:

- work standards for union employed mechanics (e.g. specification for inspections was mapped against actual inspection times);
- integration with warranty tracking, fueling, parts manuals, payroll, inventory, and purchasing modules.

The project was implemented despite an unrelated, 40-day union work stoppage. The bus fleet attained four million additional miles of service with 10% fewer mechanics—easily outweighing the $2,500,000 spent on the new system. The system became a model for the industry and was demonstrated with pride to other transportation agencies coast-to-coast.

Ask Yourself:
Will the value be transient or everlasting?

Consider the following exhibit. Technology has changed in large waves over the last 30 years such that if you didn't catch the wave you're either out of business or need to wait for the next wave . . .

Evolution of Computing Technology

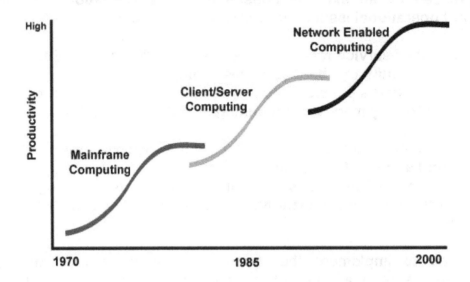

Now consider the exhibit below. Technology in the high technology sector has changed so rapidly that the notion of "wait until we're sure" may leave some companies so far behind their competitors that there existence will be threatened.

What	Late '70s	2001	2025 - ?
Personal Transport	Station Wagon	Minivan, SUV	Electric Cars; return to Mass Transit
Recorded Music	12" Vinyl LP	CD	Personal "Jukebox"
Video Recording	Almost Unknown	VCR, DVD, Video Rentals	DVD only; Personal "Jukebox"
Instant Pictures	Polaroid	Digital Camera	Real Time editing
Television Set	Rotary Controls	Pushbutton Remote	Selectors via voice
Microwave Oven	Very Unusual	Commonplace	Only way to cook
Tennis Racquets	Laminated Wood	Steel, Carbon, Titanium, Ceramic	Larger balls for slower players
Kitchen Knives	Carbon Steel	Stainless Steel, Ceramic	Ceramic
Athletic Competition	Olympic Games	X-Games	Olympics adds X-sports
Music Television	Lawrence Welk	MTV, VH-1	Personal "Jukebox"
Special Effects	Optical composition	Digital animation	Virtual reality at every desktop

Internet	Did Not Exist (ARPANET)	Ubiquitous for many	Ubiquitous for most
Networks	SNA 9.6 Kb	1500Kb (1.5 Mb)	225,000Kb (225Mb)
Dial-Up	1200 Baud	56k	May Not Exist
Recordable Media	Open Reel Tape	Cassette, CD	CD (many shapes/sizes)
PC Speed	5 MHz	2 GHZ	800 GHZ
Computer Terminal	Teletype KSR-33	Desktop PC	Voice
Input Media	Punched Card	Desktop PC	Voice
Display Monitor	Monochrome	Color CRT or LCD	LCD only
Pagers	Did not exist	Ubiquitous	May Not Exist
Mobile Telephone	Uncommon & Exotic	Ubiquitous for many	Ubiquitous for most
Document Storage	Hard copy archive	Digital Imaging	Paperless mandates
Personal Time Management	Day-Timer	Palm Pilot	Overtaken by expanded phone
Security	ID Card	Magnetic Strip Badge	Smart Card or biometrics

Some specific Internet-based trends are depicted in the following exhibit (based on a slide given to me by Dave Andrews of Oracle Corporation):

Specific Internet Trends

Administration & Finance
- Electronic Procurement
- Tax Filing
- Electronic Commerce

Health
- Telemedicine
- Teleradiology
- Claims Processing

Human Services
- Job Searching
- Integrated Case Management
- Integrated Service Delivery

Public Safety
- Handgun Registries
- Crime Statistics
- Most Wanted Lists

Courts & Criminal Justice
- CJIS
- Video Arraignment
- Lawyerless Litigation

Transportation
- Intelligent Transportation
- Traffic Flows
- Drivers License Registration

Clearly, technology changes rapidly and no one can predict any technology's longevity. All the more reason to rapidly complete IT projects the company agrees to engage.

Ask Yourself - one final question:
At the end of projects, does the company look back to determine if the effort fulfilled the hype felt at the start of the project?

This last step in project work is often overlooked. By the end of projects, team members are more relieved than celebratory, anxious to go on to something new. Yet it is important for executives to create a score card seeking answers to project budget versus actual expense, project expectations versus results, project impact on business efficiencies and/or competition, and project impact on the people of the company it effected. This review is not to evaluate the performance of the project team members or their sponsors but rather a point in time view of the company's culture for change - how did WE handle the project.

As each project is <u>scored</u> in this fashion, the company's propensity for risk associated with new technology projects is measured. It will help forecast future success or failure.

How does your company score?